XTREME AIRCRAFT

ELECTRIC AIRCRAFT

A&D Xtreme
BOLD HI-LO NONFICTION
An imprint of Abdo Publishing
abdobooks.com

S.L. HAMILTON

TAKE IT TO THE XTREME!

GET READY FOR AN XTREME ADVENTURE!
THE PAGES OF THIS BOOK WILL TAKE YOU INTO
THE THRILLING WORLD OF ELECTRIC AIRCRAFT.
WHEN YOU HAVE FINISHED READING THIS BOOK, TAKE THE
XTREME CHALLENGE ON PAGE 45 ABOUT WHAT YOU'VE LEARNED!

ABDOBOOKS.COM

Published by Abdo Publishing, a division of ABDO, PO Box 398166, Minneapolis, Minnesota 55439.

Printed in the United States of America, North Mankato, MN.

092021

012022

THIS BOOK CONTAINS RECYCLED MATERIALS

Editor: John Hamilton

Copy Editor: Tamara L. Britton

Graphic Design: Sue Hamilton

Cover Design: Laura Graphenteen

Cover Photo: Shutterstock

Interior Photos & Illustrations: AP-pgs 4-5 & 26-27; Bye Aerospace-pg 24; Deutsches Museum-pg 12 (bottom); E. Schoeberl-pg 12 (top); EADS-pg 30; Eviation-pgs 40-41; Fédération Aéronautique Internationale-pgs 8 & 9; Getty Images-pgs 18-19 & 31; H55/Anna Pizzolante-pg 25 (bottom); Lange Aviation-pgs 16 & 17; Library of Congress-pgs 6-7; Lilium-pgs 36 & 37; magniX-pgs 28-29 & 44; NASA-pgs 10-11, 38-39 & 42 (inset); Opener-pgs 34-35; Pipistrel-pg 25 (top); Rupert Composite-pgs 32-33; Shutterstock-pgs 1 & 17 (background); Solar Flight-pgs 14-15 & 20-21; Solar Impulse Foundation-pgs 22-23; Tier 1-pgs 42-43; Wikimedia-pg 13.

LIBRARY OF CONGRESS CONTROL NUMBER: 2021943571

PUBLISHER'S CATALOGING-IN-PUBLICATION DATA

Names: Hamilton, S.L., author.

Title: Electric aircraft / by S.L. Hamilton

Description: Minneapolis, Minnesota : Abdo Publishing, 2022 | Series: Xtreme aircraft | Includes online resources and index.

Identifiers: ISBN 9781532197338 (lib. bdg.) | ISBN 9781098219536 (ebook)

Subjects: LCSH: Aviation--Juvenile literature. | Electric airplanes--Juvenile literature. | Hybrid electric airplanes--Juvenile literature. | Electric vehicles--Juvenile literature.

Classification: DDC 629.1333--dc23

TABLE OF CONTENTS

ELECTRIC AIRCRAFT 4
HISTORY 6
EARLY ELECTRIC AIRCRAFT 8
MILLENIUM ELECTRIC AIRCRAFT 16
MODERN ELECTRIC AIRCRAFT 24
SUPER ULTRALIGHT ELECTRIC PLANES 30
VTOL ELECTRIC PLANES 34
CUTTING-EDGE ELECTRIC PLANES 38
ELECTRIC HELICOPTERS 42
THE FUTURE 44
XTREME CHALLENGE 45
GLOSSARY 46
ONLINE RESOURCES 47
INDEX 48

CHAPTER 1

ELECTRIC AIRCRAFT

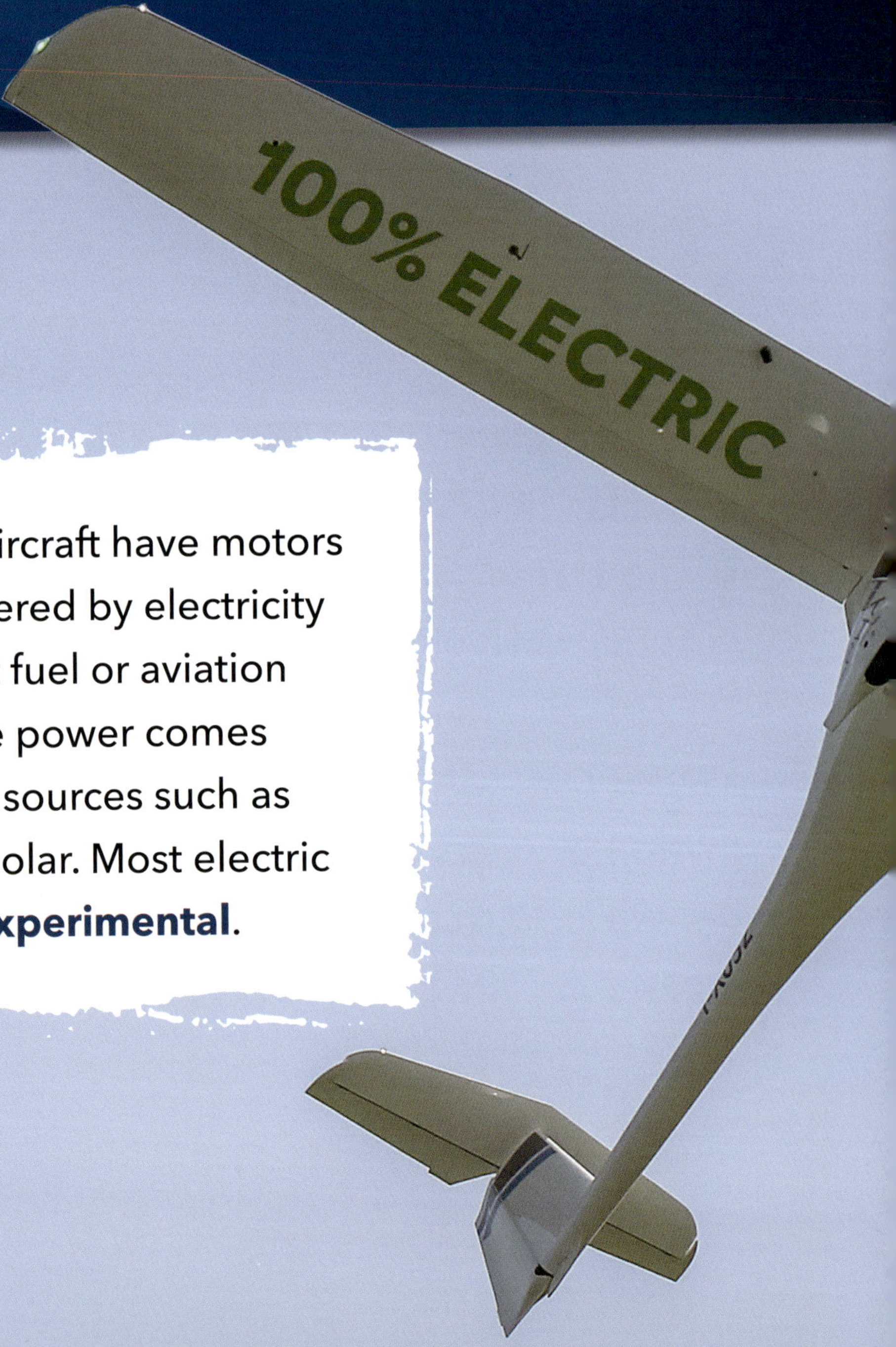

Electric aircraft have motors that are powered by electricity instead of jet fuel or aviation gasoline. The power comes from electric sources such as batteries or solar. Most electric aircraft are **experimental**.

The energy needed to fly depends on an aircraft's weight. New lightweight aircraft with powerful batteries have made electric flight a reality. Electric aircraft can travel cleanly and carry people quietly into the sky.

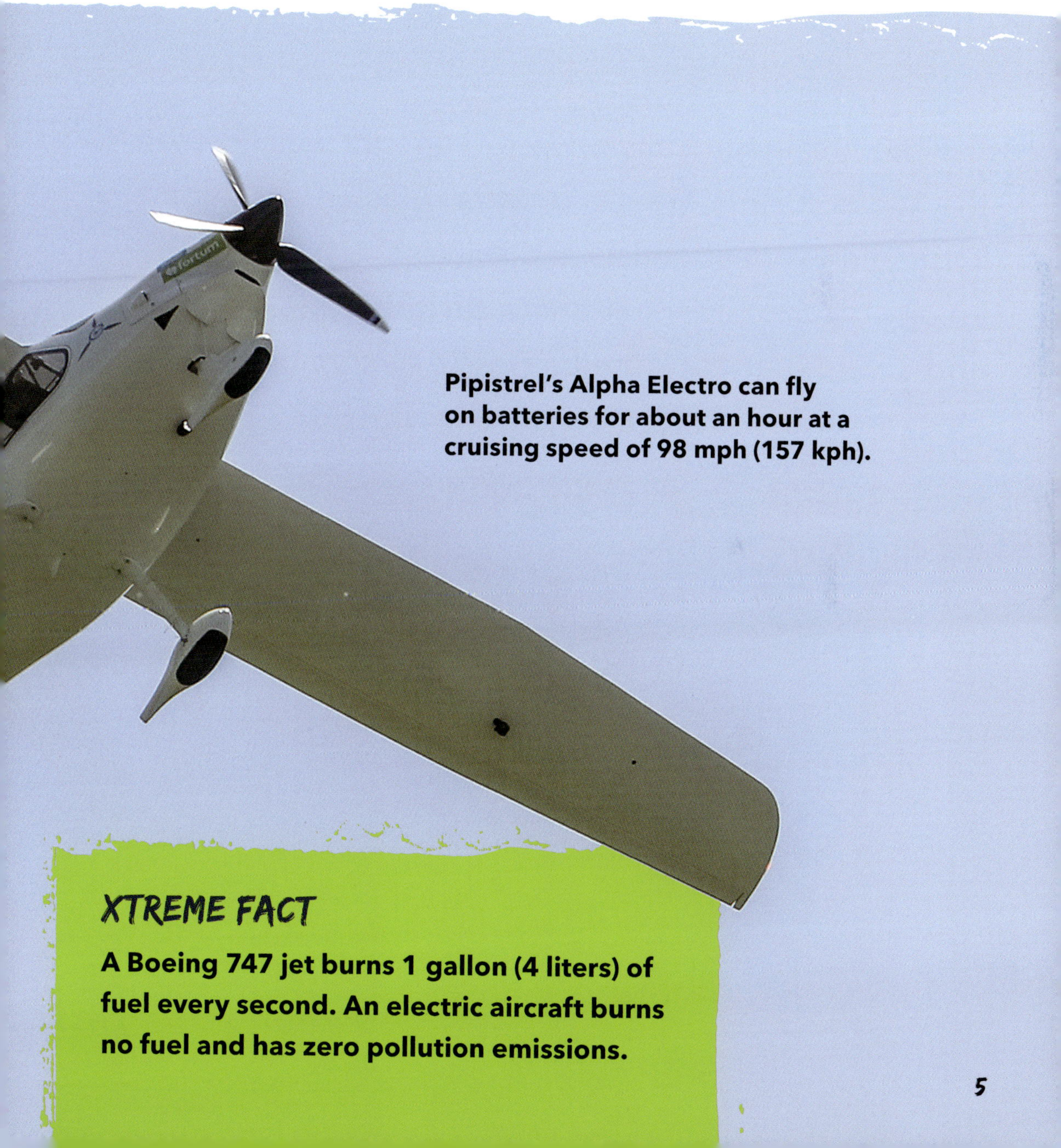

Pipistrel's Alpha Electro can fly on batteries for about an hour at a cruising speed of 98 mph (157 kph).

XTREME FACT

A Boeing 747 jet burns 1 gallon (4 liters) of fuel every second. An electric aircraft burns no fuel and has zero pollution emissions.

CHAPTER 2

HISTORY

The first electrically powered aircraft was flown in the 1880s in Paris, France. Brothers Gaston and Albert Tissandier were scientists and aviators. Gaston fitted an electric Siemens motor to a **dirigible**. He, Albert and a third man took the *La France* airship on the first electric-powered flight on October 8, 1883.

The *La France* took off and maintained steady flight in 1883, but the heavy weight of the batteries made electric power impractical.

CHAPTER 3

EARLY ELECTRIC AIRCRAFT

The Militky MB-E1 was the first electric plane to carry a person. Engineers Fred Militky and Heino Brditschka powered the MB-E1 with **nickel-cadmium (NiCad) batteries**. Brditschka piloted the electric plane for about 12 minutes on October 21, 1973, circling Wels Airfield in Wels, Austria. Additional short flights followed in the days ahead.

Heino Brditschka, Pilot

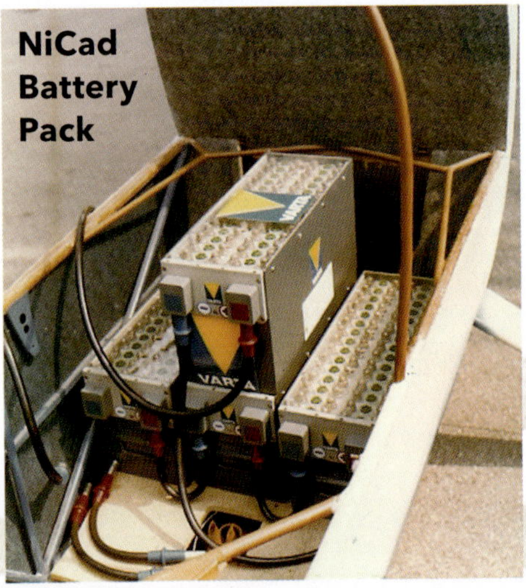

NiCad Battery Pack

The Militky MB-E1 had a wingspan of 39 feet (12 m) and weighed only 970 pounds (440 kg).

The first human-carrying solar-powered plane was the *Gossamer Penguin*, created by Paul MacCready. Power came from a solar panel mounted over the 72-foot (22-m) -long wing. To keep the weight down, its first official flight was helmed by 100-pound (45-kg) female glider pilot Janice Brown on May 18, 1979.

XTREME FACT

Early test flights of the *Gossamer Penguin* were conducted by Marshall MacCready, the creator's 13-year old son. Marshall weighed only 80 pounds (36 kg) and was an experienced pilot.

The *Gossamer Penguin* used a bicycle tow to take off.

Günter Rochelt created *Solair 1* in the early 1980s. The light glider was powered with 2,499 solar cells across its 52-foot (16-m) wings. A **NiCad** battery powered its takeoff. *Solair 1's* longest flight was 5 hours and 41 minutes, on August 21, 1983, in Germany.

***Solair 1* takes off on a test flight in 1981.**

Günter Rochelt

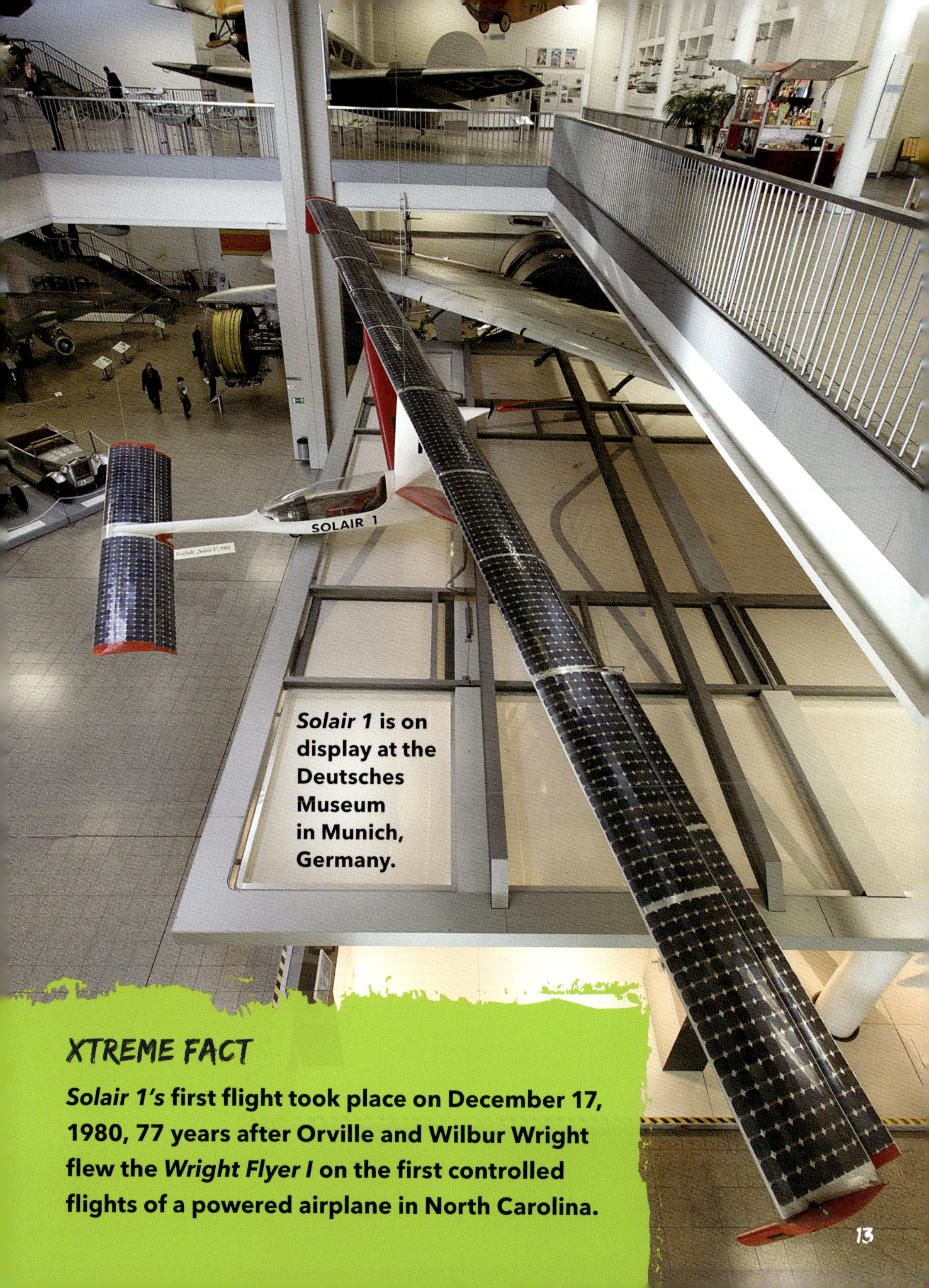

***Solair 1* is on display at the Deutsches Museum in Munich, Germany.**

XTREME FACT

***Solair 1's* first flight took place on December 17, 1980, 77 years after Orville and Wilbur Wright flew the *Wright Flyer I* on the first controlled flights of a powered airplane in North Carolina.**

Hang glider pilot Eric Raymond flew Solar Flight's Sunseeker from California to North Carolina in the summer of 1990. This was the first solar-powered plane to fly across the United States. The plane was powered by 96 **NiCad** D-cell batteries. The batteries were charged with solar panels covering the top of the plane.

XTREME FACT

Sunseeker's average flying speed was under 40 mph (64 kph). While crossing the United States, the plane traveled over about one state a day from August 2 to September 3, 1990.

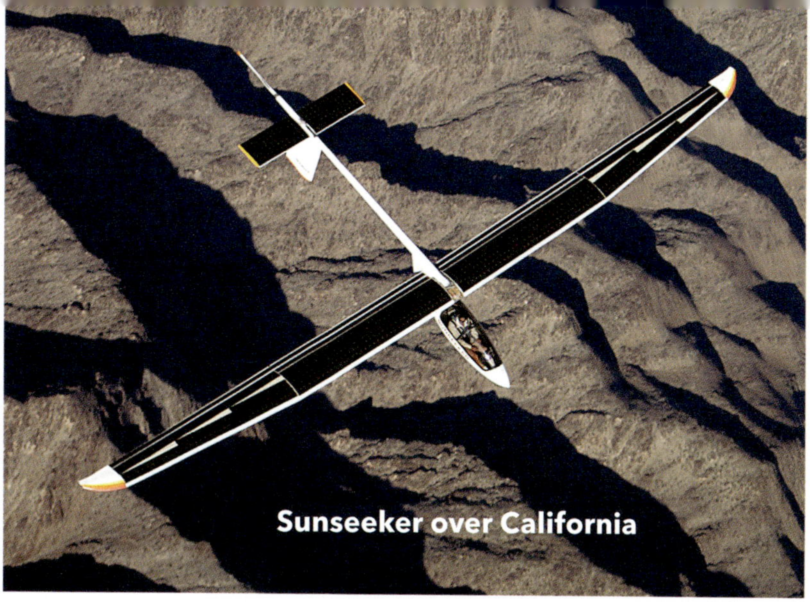

Sunseeker over California

Eric Raymond and Sunseeker's history-making coast-to-coast trip totaled 121 hours of flight time with 21 stops.

CHAPTER 4

MILLENIUM ELECTRIC AIRCRAFT

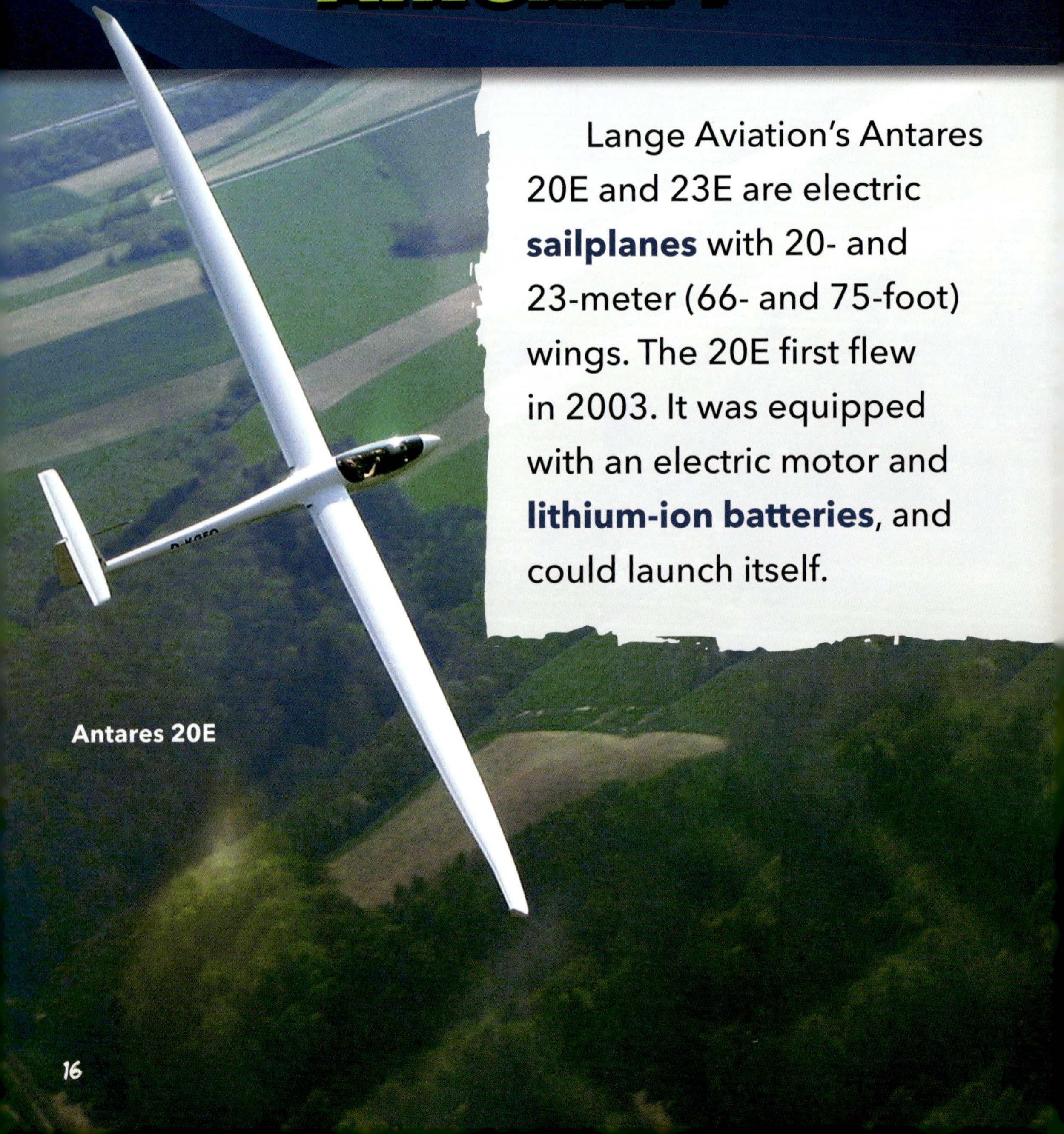

Antares 20E

Lange Aviation's Antares 20E and 23E are electric **sailplanes** with 20- and 23-meter (66- and 75-foot) wings. The 20E first flew in 2003. It was equipped with an electric motor and **lithium-ion batteries**, and could launch itself.

The Antares cockpit is a survival zone for the pilot.

XTREME FACT

The cockpit of an Antares sailplane was based on the safety design of Formula One racing cars. The nose cone absorbs energy in a crash, while the interior space is reinforced to withstand high-speed impacts.

On July 16, 2006, students at Japan's Tokyo Institute of Technology built and flew the world's first AA battery-powered plane. The single-seat glider-like plane was fitted with 160 AA Oxyride batteries. These were 1.5 times more powerful than a regular AA battery.

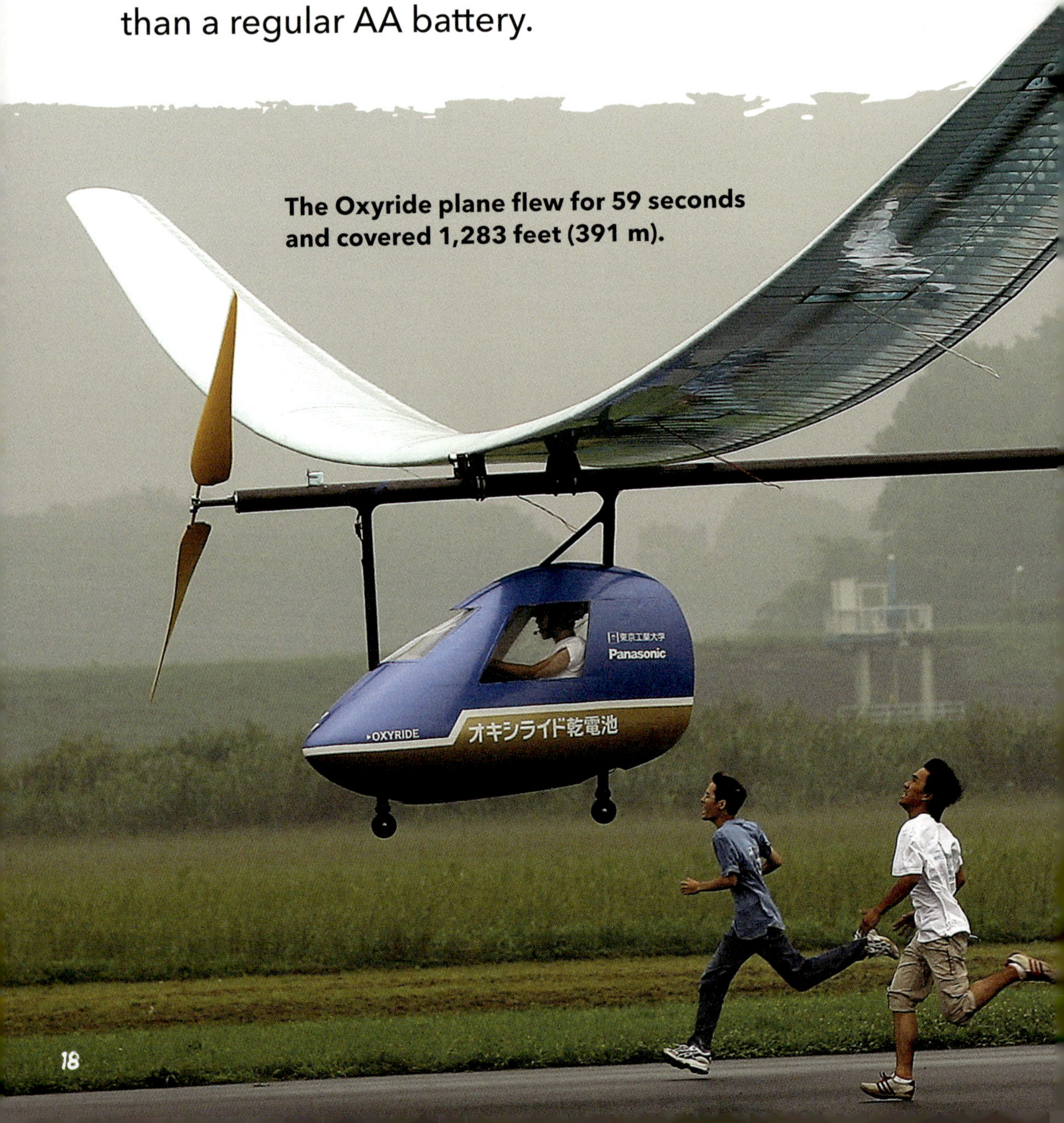

The Oxyride plane flew for 59 seconds and covered 1,283 feet (391 m).

XTREME FACT

The Oxyride plane and the pilot, Tomohir Kamiya, both weighed the same: 117 pounds (53 kg).

The Sunseeker Duo's long wings fold up, so the plane can be easily stored in a hanger.

Sunseeker Duo is a solar-powered plane able to fly two people up to 12 hours. It first flew in 2013. Its 72-foot (22-m) wingspan is covered in 1,510 solar cells that send collected energy to the battery pack stored in the **fuselage**.

The Sunseeker Duo has reclining seats and a sliding canopy that opens to allow clear in-flight views.

Switzerland's Solar Impulse 2 (SI2) became the first solar airplane to fly day and night, without any fuel, around the world. SI2's journey began in Abu Dhabi on March 9, 2015. Pilots André Borschberg and Bertrand Piccard took turns flying the 17 legs of the trip in the single-seat plane.

Solar Impulse 2 flies over California's Golden Gate Bridge in 2016. The plane had a wingspan of 236 feet (72 m) and flew at a top speed of 87 mph (140 kph).

During the flight from Japan to Hawaii, Borschberg set a record for the longest solo endurance flight of 76 hours and 45 minutes. However, battery damage due to overheating caused the SI2 to be grounded for months in Hawaii awaiting replacements. SI2 ended its 24,855-mile (40,000-km) round-the-world flight on July 26, 2016.

CHAPTER 5

MODERN ELECTRIC AIRCRAFT

Several of today's newest electric aircraft are designed to be used by flight schools to train pilots. Aircraft such as the Bye Aerospace eFlyer, Pipistrel Alpha Electro, and the H-55 Bristell Energic can fly for an hour with little noise and zero **emissions**. The **lithium-ion batteries** are changed out or quickly recharged.

Bye Aerospace eFlyer

XTREME FACT

The energy cost for an hour-long training flight in an electric plane is $3 to $7 versus $40 in a gas-powered plane.

MagniX turned a De Havilland Beaver seaplane into the world's first all-electric commercial plane. The eBeaver can carry up to nine people on sightseeing tours.

Harbour Air of Vancouver, Canada, took the eBeaver seaplane on its first official flight in December 2019.

MagniX and AeroTEC's eCaravan is a modified Cessna Grand Caravan plane capable of carrying nine passengers. It is the largest all-electric airplane to fly, making its first flight on May 28, 2020.

MagniX estimates that flying 1½ hours in an eCaravan costs $24.68 in electricity. The same flight in a gas-powered Grand Caravan would cost $404.55 in fuel.

XTREME FACT

In 2019, 49 percent of North American flights were less than 500 miles (805 km) in distance. Electric aircraft can make these flights much cheaper and with zero emissions.

CHAPTER 6

SUPER ULTRALIGHT ELECTRIC PLANES

A super ultralight plane can carry one person for a short time. Cri-Cri is a small four-engine electric airplane. It has a wingspan of only 16 feet (5 m) and a weight of just 185 lbs (84 kg). It is designed to carry the pilot for about 30 minutes, with a **cruising speed** of 68 mph (110 kph).

A Cri-Cri was the first electric aircraft to cross the English Channel. It was flown by Hugues Duval on July 9, 2015.

XTREME FACT

The Cri-Cri was designed by Michel Colomban. The engineer also helped design one of the largest planes ever built, the Concorde jet.

A small Cri-Cri sits next to a huge Airbus 380 passenger jetliner. Cri-Cri, meaning "cricket," is the nickname of designer Michel Colomban's daughter.

Archaeopteryx Electro is a special aircraft where the pilot takes off with a leap of faith. Called a foot launch, the pilot balances the craft on their shoulders, runs to the edge of a hill, and leaps off, much as a hang glider is launched.

The Archaeopteryx weighs 119 pounds (54 kg).

The pilot's legs are drawn into the cockpit and the plane's electric battery powers the flight. Archaeopteryx can fly for 11 minutes at a **cruising speed** of 36 mph (58 kph). The plane lands with regular landing gear on the bottom.

Archaeopteryx can also be launched several other ways, including with a specially fitted Electro-Drive motor that allows the craft to take off like a regular airplane.

CHAPTER 7

VTOL ELECTRIC PLANES

BlackFly is known as a personal aerial vehicle, or PAV.

XTREME FACT

BlackFly needs a distance of only 3 feet (1 m) for takeoffs and landings.

Opener's BlackFly is an electric fixed-wing **VTOL**. The unusual aircraft reaches speeds of 80 mph (129 kph). Because of its VTOL capabilities, BlackFly can take off from ground or water and can land on many different surfaces, including grass, asphalt, snow, and ice.

BlackFly in a charging station.

Lilium is an electric **VTOL** jet designed to hold six passengers and one pilot. The uniquely-shaped plane has a fixed wingspan of 46 feet (14 m) with 36 fans that rotate downward for liftoff. Its **cruising speed** is 174 mph (280 kph) and its maximum range is 155 miles (250 km).

The Lilium VTOL is able to takeoff and land on helicopter pads.

Lilium is designed to carry people and goods.

CHAPTER 8

CUTTING-EDGE ELECTRIC PLANES

NASA is working on new electric-powered flight technology to share with everyone. Their X-57 Maxwell uses rechargeable **lithium-ion** batteries for power. Two 400-pound (181-kg) battery packs are in the cabin of the 2-seater plane. It will cruise at 172 mph (277 kph) with a range of 100 miles (161 km).

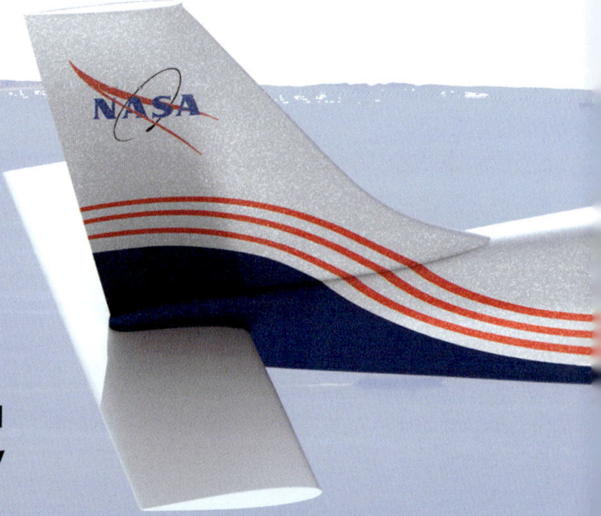

NASA hopes their technology will make electric aircraft an everyday reality. Test flights on the X-57 Maxwell are scheduled to begin in the early 2020's.

The X-57 Maxwell is a modified Italian Tecnam P2006T. To generate lift, there are 2 larger electric cruise motors and 12 smaller motors placed along the edge of the wing.

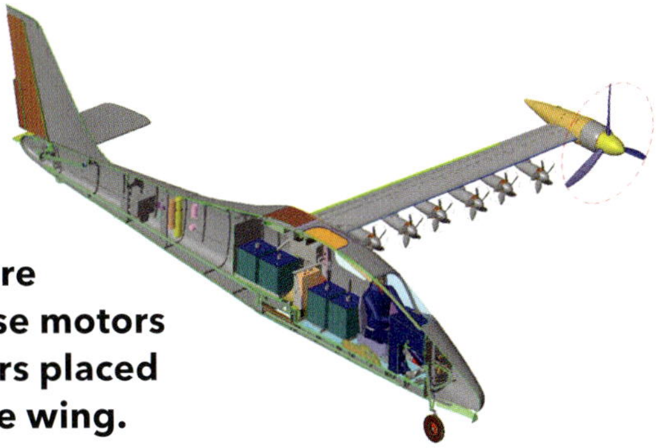

XTREME FACT

Another design for Alice had an unpressurized cabin. It would operate as an air taxi.

Eviation's Alice will be the world's first electric luxury commuter plane. It is designed to carry nine passengers and two crew members up to 850 miles (1,368 km) at a **cruising speed** of 298 mph (480 kph). The cost to operate Alice is around $200 per hour, compared to a similar size fueled aircraft at $600-1,000 per hour.

The Alice prototype was displayed in 2019, but an onboard fire delayed its production.

Alice's comfortable interior

CHAPTER 9

ELECTRIC HELICOPTERS

In 2018, Tier 1's R44 set the record for the farthest distance traveled by an electric helicopter. It flew 34 miles (55 km) at an average speed of 92 mph (148 kph).

XTREME FACT

NASA sent a camera- and sensor-carrying electric helicopter to Mars. On April 19, 2021, *Ingenuity* became the first spacecraft to achieve powered, controlled flight on another world. *Ingenuity* has a solar array on top of its rotor system that charges six lithium-ion batteries.

Tier 1 Engineering modified a Robinson R44 Raven II into a successful electric helicopter. The battery-powered copter has been designed for medical use, transporting organs between local hospitals. Tier 1's R44 has four seats. Plans are being made that would allow it to fly for 150 minutes with a payload of 600 lbs (272 kg).

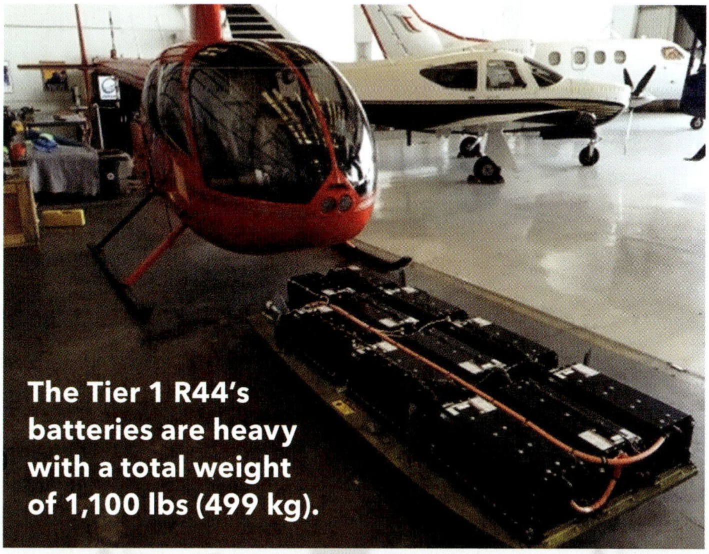

The Tier 1 R44's batteries are heavy with a total weight of 1,100 lbs (499 kg).

CHAPTER 10

THE FUTURE

Future electric aircraft will carry lighter-weight batteries that hold more power. Large electric airliners are being developed that will carry 150-175 passengers and fly at **subsonic** speeds. With improved technology, electric aircraft will be able to go faster and farther. Electric planes will continue to be quiet, reduce the use of fuel, and be more environmentally friendly.

Technology is being developed to make electric planes of all sizes.

XTREME CHALLENGE

TAKE THE QUIZ BELOW AND PUT WHAT YOU'VE LEARNED TO THE TEST!

1) Name two sources that can power an electric aircraft.

2) Are electric aircraft noisy?

3) What are the names of the brothers who flew the first electrically powered aircraft? From what country did they take off?

4) What was the first electric airplane to carry a person?

5) What was the first human-carrying solar-powered plane?

6) Electric aircraft use what types of batteries?

7) What Swiss aircraft became the first electric plane to fly day and night, without any fuel, around the world?

8) What is the name of NASA's solar-powered helicopter that flew on Mars?

GLOSSARY

cruising speed – A comfortable speed for an aircraft that is usually below its maximum, but economical fuel-wise.

dirigible – A lighter-than-air airship with a rigid frame. It was filled with hydrogen or helium to provide lift. Sometimes referred to as a Zeppelin.

emissions – Aircraft engines that burn fossil fuels for power produce emissions similar to car engines. These pollutants cause environmental concerns because of their effect on air quality. Electric aircraft have no polluting emissions.

experimental aircraft – Planes that have not been fully proven in flight. These are usually aircraft that are using new technology or aircraft that are built at home. In the US, the Federal Aviation Administration gives special airworthiness certificates within the experimental category.

fuselage – The main body of an aircraft. The fuselage usually holds the pilot, crew, passengers, and cargo.

lithium-ion batteries – Lightweight, rechargeable batteries that use a lithium-ion solution to provide stored power.

nickel-cadmium (NiCad) batteries – A rechargeable battery that uses the metals nickel and cadmium to store power.

sailplane – A glider designed to fly and gain altitude from natural forces, such as rising warm air. An electric sailplane uses some type of electricity to help power it.

subsonic – Speeds less than the speed of sound (700 mph/ 1,127 kph) or Mach 1. All civilian aircraft, and many military crafts, fly at subsonic speeds.

VTOL – Vertical takeoff and landing. An aircraft, usually other than a helicopter, that can takeoff and land vertically, usually by using rotors or propellers that tilt up and down.

ONLINE RESOURCES

Booklinks
NONFICTION NETWORK
FREE! ONLINE NONFICTION RESOURCES

To learn more about electric aircraft, please visit **abdobooklinks.com** or scan this QR code. These links are routinely monitored and updated to provide the most current information available.

INDEX

A
Abu Dhabi, 22
AeroTEC, 28
Airbus 380, 31
Alice, 40, 41
Alpha Electro, 5, 24, 25
Antares 20E, 16, 17
Antares 23E, 16, 17
Archaeopteryx Electro, 32, 33
Austria, 8

B
Beaver seaplane, 26
BlackFly, 34, 35
Boeing 747, 5
Borschberg, André, 22, 23
Brditschka, Heino, 8
Bristell Energic, 24, 25
Brown, Janice, 10
Bye Aerospace, 24

C
California, 14, 15, 22
Canada, 27
Cessna, 28
Colomban, Michel, 31
Concorde jet, 31
Cri-Cri, 30, 31

D
De Havilland, 27
Deutsches Museum, 13
Duval, Hugues, 30

E
eBeaver seaplane, 26, 27
eCaravan, 28
eFlyer, 24
Electro-Drive motor, 33
English Channel, 30
Eviation, 40

F
Formula One racing cars, 17
France, 6

G
Germany, 12, 13
Golden Gate Bridge, 22
Gossamer Penguin, 10, 11
Grand Caravan, 28

H
Harbour Air, 27
Hawaii, 23
H-55, 24, 25

I
Ingenuity, 42

J
Japan, 18, 23

K
Kamiya, Tomohir, 19

L
Lange Aviation, 16
Le France, 6, 7
Lilium, 36, 37

M
MacCready, Marshall, 10
MacCready, Paul, 10
magniX, 26, 28
Mars, 42
Militky, Fred, 8
Militky MB-E1, 8, 9
Munich, Germany, 13

N
NASA, 38, 42
North America, 29
North Carolina, 13, 14

O
Opener, 35
Oxyride, 18, 19

P
Paris, France, 6
Piccard, Bertrand, 22
Pipistrel, 5, 24, 25

R
Raymond, Eric, 14, 15
R44 (electric helicopter), 42, 43
R44 Raven II, 43
Robinson, 43
Rochelt, Günter, 12

S
Siemens, 6
Solair 1, 12, 13
Solar Flight, 14
Solar Impulse 2 (SI2), 22, 23
Sunseeker, 14, 15
Sunseeker Duo, 20, 21
Switzerland, 22

T
Tecnam P2006T, 39
Tier 1 Engineering, 42, 43
Tissandier, Albert, 6
Tissandier, Gaston, 6
Tokyo Institute of Technology, 18

U
United States, 14

V
Vancouver, Canada, 27

W
Wels, Austria, 8
Wels Airfield, 8
Wright, Orville, 13
Wright, Wilbur, 13
Wright Flyer I, 13

X
X-57 Maxwell, 38, 39